MW00800570

LIFEVIEWS

Published by Creative Education
123 South Broad Street, Mankato, Minnesota 56001
Creative Education is an imprint of The Creative Company

Art direction by Rita Marshall; Production design by The Design Lab/Kathy Petelinsek
Photographs by KAC Productions (Kathy Adams Clark, Larry Ditto, Peter Gottschling,
Glen Hayes, Greg Lasley, John & Gloria Tveten)

Library of Congress Cataloging-in-Publication Data

Rotter, Charles.
The prairie : an enduring spirit / by Charles Rotter.
p. cm. — (LifeViews)
ISBN 1-58341-029-5
1. Prairies—North America—Juvenile literature. 2. Prairie ecology—North America—Juvenile literature.
3. Prairie conservation—North America—Juvenile literature. [1. Grasslands. 2. Prairies.] I. Title. II. Series.
QH102 .R67 2001
577.4'4'097—dc21 00-045167

First Edition

2 4 6 8 9 7 5 3 1

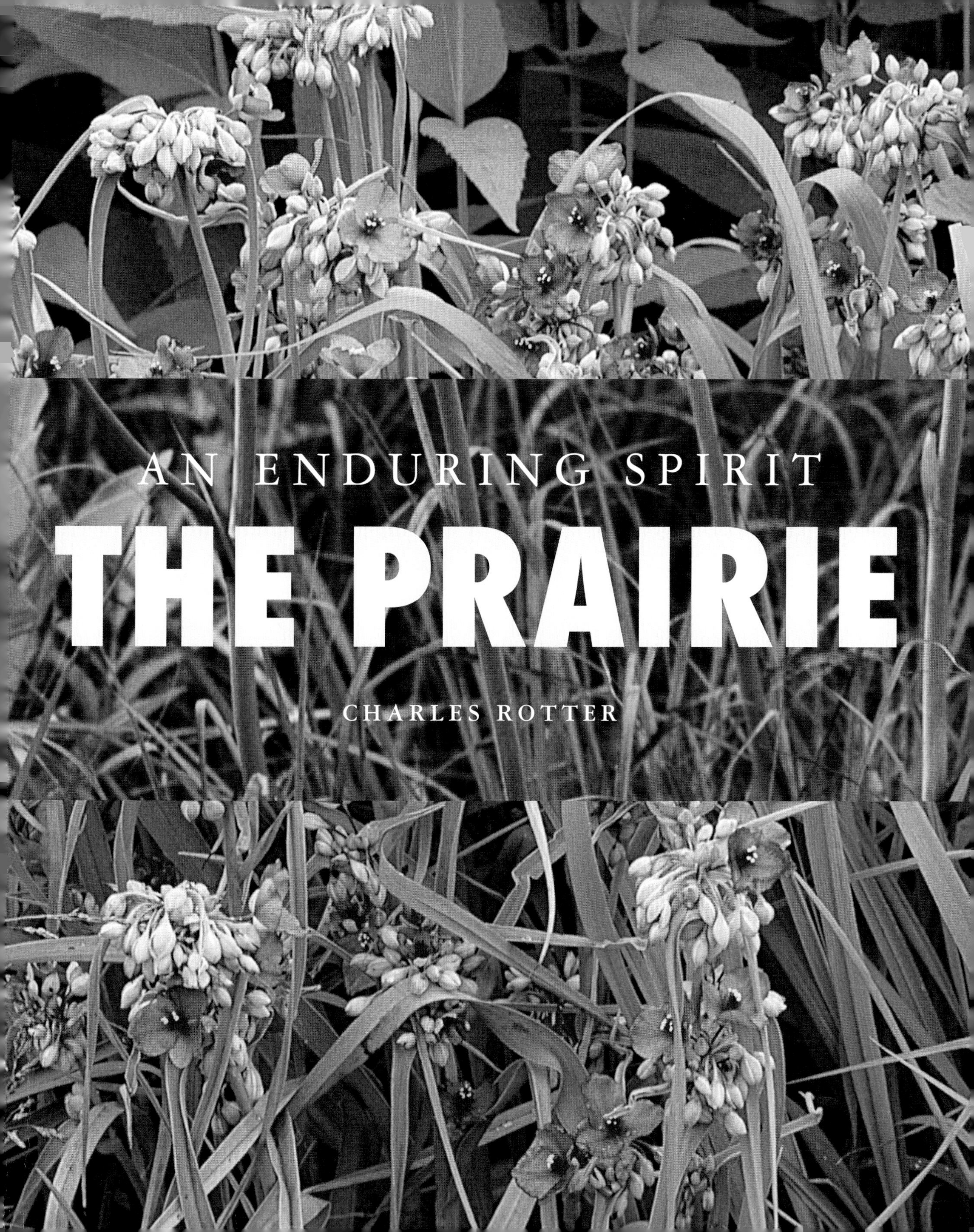

AN ENDURING SPIRIT

THE PRAIRIE

CHARLES ROTTER

PEOPLE often describe prairies or grasslands by listing how they differ from other regions. For example, grasslands are flatter and drier than mountainous areas. They are less densely overgrown than jungles, and a bit more humid and hospitable than deserts. Describing what **grasslands** are not, rather than what they are, however, does them a great disservice.

Grasslands are vigorous regions filled with many forms of life competing for survival. They are among the most fertile lands of our planet and play an important role in the global **ecosystem**. Agriculture established in former grasslands accounts for about 70 percent of the food produced in the world.

The prairie ecosystem is both fragile and beautiful.

Most grasslands are found in the middle of continents. They are far from the climate-moderating effect of oceans, so they are subject to wide daily and seasonal fluctuations in temperature and **precipitation**. Grasslands receive about 10 to 40 inches (25–102 cm) of precipitation per year. If they get less precipitation, they are likely to become deserts; more, and they are likely to become forests.

Grasslands make up more than 30 percent of the earth's land area. The world's main grasslands are: the Eurasian steppe, which extends from Mongolia in Asia all the way into eastern Europe; the African savannas; the llanos of Brazil; the pampas of Uruguay and Argentina; large tracts of frozen **tundra** throughout the arctic latitudes; and the North American prairie.

The first Europeans to encounter the North American prairie were Spanish **conquistadors** led by Francisco Vásquez de Coronado in 1541. They had been riding north in search of gold, and instead found a seemingly endless sea of grass. More than a century later, French explorers also visited the region.

Grasslands take different shapes in different parts of the world. Many people know them as fields of swaying cane (top), but grasslands also include tundras (bottom).

They named the vast grasslands with the French word for meadow: **prairie**.

We still use the word prairie to describe the grasslands spanning the great plains of North America. The prairie forms a large triangle in the middle of the continent. Bordered by the Rocky Mountains on the west side, it stretches more than 2,000 miles (3,218 km) from the Northwest Territory of Canada to the U.S. state of Texas. The other two sides of the triangle connect near the eastern edge of Iowa.

North American **prairie grasses** are divided into three general types according to their heights: shortgrasses, midgrasses, and tallgrasses. Shortgrasses, which grow up to 18 inches (46 cm) high, prevail in the West. Common prairie shortgrasses include grama grasses and buffalo grass. Midgrasses, which grow between two and four feet (61–122 cm) high, dominate in the geographic center of the prairie. Needlegrasses, little bluestem, and foxtail barley are examples of midgrasses. Tallgrasses, which usually grow more than five feet (1.5 m) and sometimes as high as 12 feet (3.7 m), are most

Pronghorns are among the largest inhabitants of the North American prairie. They are the fastest land animals on the continent and range far to feed on grass and small shrubs.

common throughout the eastern prairie. Common tallgrass varieties are big bluestem, cordgrass, and Indian grass.

The reason these grasses grow where they do is simple—shortgrasses need less rainfall than midgrasses, which in turn need less than tallgrasses. The total annual rainfall on the prairie increases steadily from west to east. This increase is caused by the Rocky Mountains. As air from the West Coast rises over the mountains, it cools, dropping much of its moisture as rain or snow on the windward side. This creates a region of little rainfall—known as a **rain shadow**—on the leeward side of the mountain. As the air moves east, it is supplied with moisture from other sources to the north and south, and the amount of rainfall increases.

Because the eastern prairie gets more precipitation than the prairie to the west, it supports a wider variety of plants and animals. Hundreds of different grasses can exist within a single acre (.4 hectare). The soils are rich black **topsoils** called mollisols. The mollisols of the eastern prairie are among the most fertile and productive in the world. Much of North

Grassland vegetation varies widely. Many areas of the African savannas are characterized by thin grasses (top), while large sections of the North American prairie feature weedy herbs such as goldenrod (bottom).

America's agriculture, particularly corn, wheat, and soybean production, has developed in lands of the eastern prairie.

Grasses are ideally suited for life on the prairie. They can survive drought, freezing temperatures, drying winds, and grazing animals—conditions that would kill many other plants. The grass grows thick and dense, forming a tough layer of matted soil called **sod**. A thick layer of sod can keep the roots from freezing in the winter and even insulate the underground part of the plant from the heat of a **brushfire**. Prairie grasses also have deep and efficient root systems that can seek out moisture located far underground. And because grass grows from the base of a blade, it continues to grow if the blade is cut.

Alongside the many different types of grasses growing on the prairie are hundreds of broad-leaved herbs called **forbs**. Some common prairie forbs are fleabane, milkweed, and gentian. Many of these herbs were used as medicines by both Native Americans and European settlers. For example, the Omaha Indians used milkweed roots to make a salve for

Prairie brushfires enable new grass to grow.

wounds. The European settlers used the same roots to treat respiratory ailments. Together, the forbs and grasses form the heart of the prairie ecosystem. They are the base of the **food chain** upon which all other forms of prairie life depend.

Before Europeans imported horses, cattle, and sheep, only two large grass-eating animals roamed the prairie—the mighty **bison** and the swift pronghorn. While these large mammals eat tremendous amounts of grass, they actually help to maintain a healthy supply of grass as well. By clearing away **old growth** while feeding and by breaking up the sod with their sharp hooves, these animals inadvertently till the soil, promoting the growth of new and healthy plants. The animals' droppings are also an excellent source of **nutrients** for the growing plants.

Native Americans hunted for centuries without depleting the bison, which numbered in the tens of millions. White hunters in the 19th century, however, killed the animals ruthlessly. Native Americans hunted only out of need, but the new

Bison (top) have long been a symbol of the North American prairie, even though their population is very limited today. Smaller animals such as bobcats and rabbits (bottom) are more widely spread.

arrivals hunted for sport and profit, slaughtering the herds until the bison was nearly extinct. Although the bison have made a comeback in population, they now live mainly in refuges and on ranches. Never again will they darken the prairies from horizon to horizon.

Predators such as wolves and coyotes have also long roamed the prairie. The bison's size, strength, and habit of clustering in large herds help protect it from these predators. The **pronghorn**, however, has different means of defense. It has keen eyesight to spot danger at a distance and is one of the fastest animals in the world. Unlike many grazing animals, which run back and forth when pursued, pronghorns run in straight lines to simply outdistance their enemies.

In addition to pronghorns and bison, many **rodents** live off the prairie grasses. Many of these small, burrowing mammals are social animals and may live in large, cooperative communities. The prairie dog is the most social of the rodents, living in clans called coteries. The coteries often include more than a dozen prairie dogs who feed, play, and groom together.

Rodents and the predators that hunt them make up much of the wildlife on the prairie. Common among these animals are prairie dogs (top), mice, rattlesnakes, and jackrabbits.

Members of a coterie are usually very protective of their **territory**, but in times of danger, they allow other prairie dogs to use their burrows to hide from predators such as hawks or coyotes. Sometimes the coteries even live side by side, forming the equivalent of towns or even cities. At the turn of the century, for example, a community of prairie dogs in Texas reportedly covered 25,000 acres (10,125 hectares) and contained 400 million prairie dogs.

The birds of the prairie are also uniquely qualified to survive in the grasslands. Most of them need less water, eat more seeds, and fly together in groups more often than birds in other climates. Some of them, such as the prairie chicken, are ground dwellers, spending most of their time roaming on foot in search of food.

While mammals and birds may be the most noticeable animal life on the prairie, they are far outnumbered by smaller animals that are much harder to see. Most of the creatures that inhabit the prairie live deep within the

Wild turkeys (top) and prairie chickens (bottom) are among the largest birds native to the prairie. These birds tend to move in flocks, especially in the winter.

grass and soil. Insects and unsegmented worms called **nematodes** consume far more grass and plants than all the mammals and birds put together. Insects, worms, nematodes, fungi, microscopic bacteria, and **protozoans** are all part of the soil ecosystem. Some benefit the grasses; others do not. But almost all of them play some crucial role in stabilizing the circle of life on the prairie.

Occasionally, however, the prairie may seem anything but stable. For example, **locusts**, a type of grasshopper, sometimes multiply out of control. When this happens, billions of locusts cover the sky like clouds, flying in search of food. They strip all vegetation in their path, destroying crops and leaving only barren fields behind them. Eventually, the swarm dies out because of a lack of food, a change in climate, or an increase in its natural enemies, such as birds. Fortunately, because the prairie grasses have such large underground **root systems**, they can grow back and replenish the plains after even the worst locust attack.

In this situation and others, the checks and balances of nature ensure that the ecosystem will not be permanently

Few prairie insects are as numerous or visible as grasshoppers. Although all grasshoppers feed on plants, most except for locusts do little harm to farmers' crops.

damaged by changing conditions. When people use the grasslands, however, these checks and balances often disappear. Unfortunately, grasslands that have been converted to agriculture may no longer be as resilient as they once were. Without the protective cover of sod, winds can carry off tons of precious topsoil, robbing the land of its productivity.

This is what happened in the United States during the dust bowl in the 1930s. Prolonged **drought** killed the crops, and because the sod had been plowed away, the dry winds stripped the topsoil off the farmland, filling the skies with clouds of dust. It was among the worst agricultural and environmental disasters in U.S. history. Some of the dust drifts piled up to 25 feet (7.6 m) high. In several states, the amount of dust blown into the atmosphere was enough to make people seriously ill.

Soil **conservation** methods, including terracing, windbreaks, and controlled grazing, can help protect the grasslands. But these practices are often expensive and may reduce farming profits. Because of this, many of the world's farmers—who are often struggling financially—do not use these conservation

Grasslands are home to many beautiful flowering plants. But prairie flowers offer more than just color. They also provide food and habitat for many spiders, butterflies, and other insects.

methods. Eventually, new grassland is cleared to replace the ruined crop land, and the destructive cycle continues.

In African grasslands, the main problem is **overgrazing**. When domesticated animals are allowed to graze too long in one area, the root systems of the plants become damaged, and the soil erodes easily in dry weather. Overgrazing has contributed to the southerly advance of the Sahara Desert over the past 5,000 years. While climate changes are mainly responsible for this loss of more than 250,000 square miles (647,500 sq km) of productive land, misuse has certainly accelerated this process. As the land becomes barren and people move on, all that is left is an unproductive desert landscape. This process is known as **desertification**.

As the world's population continues to grow, so does the pressure on grasslands throughout the world. The people who most need land to grow crops and raise animals are the ones who will suffer later if conservation is not practiced. It is up to the people of the entire world to understand, appreciate, and protect the fragile **environment** of the world's grasslands.

Overgrazing and erosion can cause irreparable damage to grasslands.

ACTIVITY ONE

LOOKING AT SOIL

Sod is critical in sustaining life on the prairie. This dense layer of soil holds the roots of plants and grasses, forming a tough mat that gives the land a kind of protective barrier. Beneath the sod are other layers of soil that cannot support plant life as well. You can study these different soil layers in your own backyard.

You Will Need

- A soil sample with grass growing from it (be sure to take it from a place where you have permission to dig)
- A garden spade
- A piece of cardboard
- A ruler

Collecting the Soil

1. Find a spot where the grass is green and grows thick. Don't worry about ruining the grass. You can replace the soil and grass when you are finished.
2. Use the spade to cut out a sample of soil with the grass intact. The sample should be about four inches (10 cm) deep, two inches (5 cm) wide, and four inches (10 cm) long. To cut the sample out of the ground, stick the spade straight down into the soil and pull it out again. Do this on all four sides of the sample. When you insert the spade along the fourth side, pull back on the handle to loosen the soil. Use the spade to lift the sample out of the ground, then place it on its side on the piece of cardboard.

Observation

1. Look at the sample, making note of the different layers of soil.
2. First, look at the top layer, which has the grass growing in it. This layer is the topsoil. Take some of the topsoil in your hand and squeeze it. It should be loose and come apart easily. It will probably feel a little damp, since this layer of soil holds moisture well. Topsoil is the best kind of soil for growing plants. It includes humus—a material made up of decayed plants and animals—that helps to stimulate plant growth.
3. With the ruler, measure the depth of the topsoil. Also examine its color. Generally, the darker the color, the better it will support plant life.
4. Now look at the next layer down. This layer, called subsoil, may consist of either clay or sand and be gray or reddish in color. Take some of the subsoil and squeeze it. It should stick together. If it's wet, it will feel slippery; if it's dry, it will feel hard and packed. Clay soil holds water for a long time, and sandy soil allows water to flow through very quickly. Neither type could support plants well without the topsoil. Roots do not grow well in clay soil, and they may not find enough water in sandy soil.
5. When you are finished studying your soil sample, put it back into the ground where you found it. Gently press it into place with your foot.

Because your sample was growing thick green grass, you know that it was good soil. But how good is the soil in other places, such as on hillsides or river banks? You may want to study other soil samples to make comparisons. How does the soil in these places differ in layer thickness, color, and feel? Does its grass grow better or worse than the grass in your first sample?

ACTIVITY TWO

LOOKING AT ROOTS

Roots anchor grass and plants to the soil, allowing them to grow. But roots also help the soil by protecting it from wind and rain, which can cause erosion. To see this for yourself, you will need a plant that has been growing for a while in a pot. Be sure to get permission to use the plant for the experiment. Grasp the plant firmly near the bottom of the stem and carefully dump it out of the pot. Look at the roots.

The soil will cling tightly in a clump among the roots, with little falling off. This happens because, as the plant grows, its roots extend into the ground. As the roots anchor the plant to the ground, they also lock the soil particles together so they are not washed away. If the plant were to die, the soil would lose the protection of the roots and be more vulnerable to erosion.

LEARN MORE ABOUT THE PRAIRIE

Conservation Alliance of the Great Plains
P.O. Box 22809
Lincoln, NE 68542
http://www.conservationalliance.org/

Grand Prairie Friends
P.O. Box 36
Urbana, IL 61803
http://www.prairienet.org/gpf/homepage.html

Iowa Prairie Network
1308 160th Avenue
Knoxville, IA 50138
http://www.IowaPrairieNetwork.org/

The Missouri Prairie Foundation
P.O. Box 200
Columbia, MO 65205
http://www.moprairie.org/

National Grasslands Visitor Center
708 Main Street, P.O. Box 425
Wall, SD 57790
http://www.fs.fed.us/grasslands/loc.htm

Neal Smith National Wildlife Refuge
Prairie Learning Center
9981 Pacific Street, P.O. Box 399
Prairie City, IA 50288
http://www.tallgrass.org/

INDEX

Both wild prairie and agricultural land need to be carefully protected.